Publication:

Natural Time Measurement

A Unique and Universal Approach

by Rinon Hoxha,

servant of the Eternal, only.

Date 14, Month 12 of the Natural Year

(16.12.2013)

~ Preface

Author belongs only to God alone, and to no religion, hence is free to notice the word and meanings of the God without any doctrine or tradition. Also, since this is an open theme, whoever has to shed more light into this topic, is encouraged and welcomed to do so. The approach to this study was sensical, intuitive, inspirational and logically argumented with the word of God throughout holy scriptures.

The conclusions followed the findings.

Peace be upon you and all this work is dedicated to God, the One who made all this possible.

Table of Contents

~ Intro

Had you been solo in the woods, with nothing but you, and holy scriptures by the side, how would you measure time?

Given you know how to read and count.

And not just for a day, but at least for 19 years. With no internet, no mobile, no watch to watch. Just your eyes, your intelligence, the Presence of God, and His word.

?

First observations you would make is the birth of sun each dawn of the day. And the sunset at the beginning of each night. So night following the day. Some clear nights would reveal the moon, and the moon is not the same each night. It appears in stages of light, from darkness to complete insight.

So here are: The sun and the moon, heavenly bodies uninfluenced by the man (thank God) to count.

Always precise from the days of genesis to the day of judgment.

Ask anyone these days, what is the purpose of the moon besides its beauty, and you'll get enough blank looks.

Now learning to judge first requires learning to see.

From the woods, the moon is seen repeating itself. Sometimes each 29, sometimes each 30 days.

And the seasons repeat itself each 12 and sometimes 13 moons.

Since the beginning of creation was dark, it makes sense that the beginning of the moon-th is dark moon, empty and ready to be filled with light.
So, zero moon/new moon marks the beginning of the new month.

Noticing seasons at first seems that the year begins in spring, when the nature blooms, yet not then. It begins when the nature is ready to begin to bloom. Like the zero moon.

So at zero season is the time to begin the first moon of the year.

It is, thus, winter. The time when snow purifies the pollution of the past year and the cycle is ready to begin again.

Being in nature, allows to see that the days and the nights are different in length during each season. In winter when the natural year begins, the nights are longest. As the year travels to more light during the day, the nights get shorter, till a point when they are equal, hence spring equinox.

This is the first turn of the year. From longer nights to longer days. Then the days continue to grow and the nights to shorten until the point of summer when the day is longest and the night the shortest. This is called summer solstice. It is the second turn, hence the half of the year.

Then nights begin to get back to where they started, thus first arriving to get equal with the day, arriving at autumn equinox, the third turn of the year, and continuing to the winter solstice, where they began as the longest nights. With this a complete cycle of seasons is rounded. And the natural year begins again.

The moon has it's own turns as well. The main being: Half moon, full moon, the second half, and Zero moon. The moon is mirror of the sun, since it reflects the sunlight.

The year begins at zero season, winter.

The moon begins with zero light, beginning of the month.

The year's first turn is in the spring equinox, when the daylight becomes equal with the nightdarkness.

The moon's first turn is at the first seven, when the first half of it's light is equal with the complement of her darkness.

The year's second turn is in the summer solstice, when the daylights are longest.

The moon's second turn is at the second seven, when it is full of light.

The year's third turn is in the autumn equinox, when the nightdarkness becomes equal with the daylight.

The moon's third turn is at the third seven, when her first half of the dark becomes equal with the complement of her light.

The year's fourth turn is back to zero, in the winter solstice, when the nightdarknsess is longest.

The moon's fourth turn is at the fourth seven, when it goes towards zero/dark moon.

And everything begins from the beginning again. Thus the moon mirrors the seasons of the year, which are dependant on the sun. Peace, and the praise to the Lord of all grace.

~ Names or Numbers of the Days?

If you choose to judge with the word of God, you will notice that in Genesis, there are no names for the days. Todays implemented Gregorian calendar has derived the names according to the planets, moon-day, sun-day, saturn-day. And in my albanian language: Mercury-day, for Wednesday.

However these names are another form of planetary idolatry.

In Genesis we have numbers: 1st to 7th day of Creation.

Praise be to the One whose Name is Eternal.

~ Beginning of the Week and the Month

When you read 12th chapter of Exodus, God commands the children of Israel to take a lamb in the 10th day of the first month and sacrifice it at dusk of the 14th day, because at night after that day the Lord would pass over Egypt to cause a final plague.

Isaiah - *66:23 From one New Moon to another and from one resting day to another, all mankind will come and bow down before me," says the Lord.*

From this verse, new moon days and resting days are non-working days.

Regarding the count of months, here are two more convergent verses from Koran and Torah:

Koran - *9:36 The count of the months with God is twelve months in the book of God since the day He created the heavens and the earth;*

Torah, Genesis - *2:4 These are the generations of the heaven and of the earth when they were created, since the day that the Lord God made earth and heaven.*

In the natural month, the month and the week do not start at the same day.

The month begins with the new moon, which is counted as the first date.

When one says the first day of the week, that's the the first day as God named it in Genesis, a working day. While the first day of the month is one day before it, the new moon day.

So, the first day of the month is the date one.

And the first day of the first week is the date two.

And so it goes until the end of the month. Also, there are many months/almost each second month during the year which at their end have "slider days" before the moon completing the cycle around the earth, which is one day after the last week and before the new moon day.

The date is the sum of the day and the night.

Praise be to God. The true insight.

The moon is very hard to be noticed in the new moon day, since it is very very thin since it is the beginning of her receiving the light. Whereas in the first day of the week, the second day of the month, the crescent moon is clearly visible for every eye. And so it grows in light until the middle of the month.

Then, in Exodus God has chosen the night after the day 14 of the first month to act, and the dusk of that day for te people to sacrifice the lamb. The night before the second resting day, or the night of date 14. When the moon gets completely full.

Naturally, the full moon today is the time when the living is most active. And the Lord swears by the full moon in Koran.

Praise to Him.

Peace.

~ *The Cycle of a Month*

The truth is plain. So we'll talk plainly:

From now on we're going to use the Equidistant Azimuthal Map to explain the movements of the sun and moon above the plain true earth.[1]

For quick difference, in the encircled plain earth, there are no hemispheres, but various perimeters/circumferences upon which the sun and the moon float respectively above in the own heavenly dome.

The second main difference is that the North is the centre of the plain true earth and the south is the last and broadest perimeter of Antarctica.

Equator is the middle circumference/perimeter of the cirlce plain true earth.

For counting months is responsible the moon. It by design changes the major phases each heavenly 7 days to let people know at which part of the month we actually are.

However, there is a dance in the floating between the sun and the moon, the two luminaries dedicated for counting time.

Let's see first the diagram representing the beginning of the month:

---

[1] Check the Publication "The Scriptural Heaven and Earth".

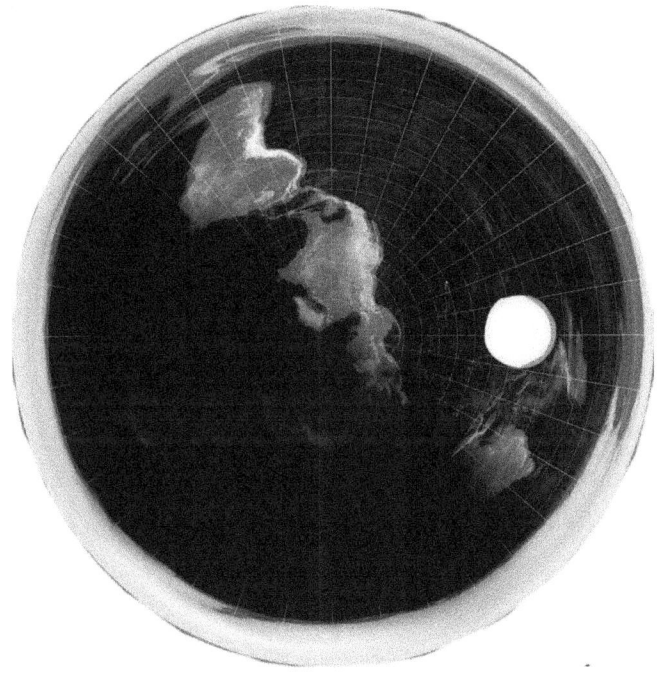

The month begins when the moon is not seen.[2] At zero light.

"At **New Moon**, the Moon rises in the morning; it's at its highest, in the south, in the middle of the day and it sets in the evening - just like the Sun."[3]

As you can see from the above representation, both luminaries are on the same spot, moving clockwise around the North pole, which is the center of the pie. The sun spins around the seasonal perimeter by one cycle during 24 hours.

The moon however begins to create distance with the sun in its birth and setting, as her phases show up from the first crescent towards the first quarter.

---

[2] Please check the Annex 9 "The Final Revelation for beginning of Moon-th".
[3] From the article – What is the Moon doing?

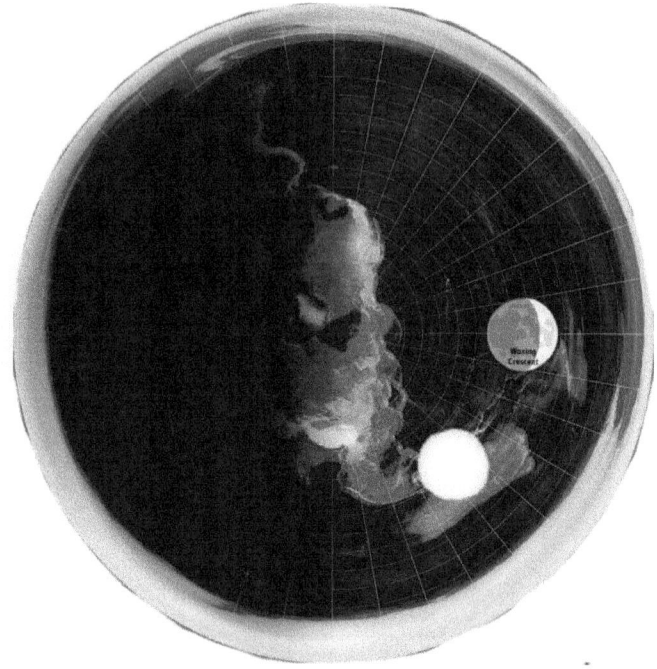

This is the beginning cycle of the month when the moon illuminated by itself (and not by reflection of the sun) shows that the first week is ongoing, until the first quarter.

"By **First Quarter**, the Moon is one-quarter of the way around its orbit (and half illuminated). It is now 90 degrees to the left of the Sun, and lags behind it by 6 hours. So it rises in the middle of the day, it's high in the south at sunset, and it sets in the middle of the night."[4]

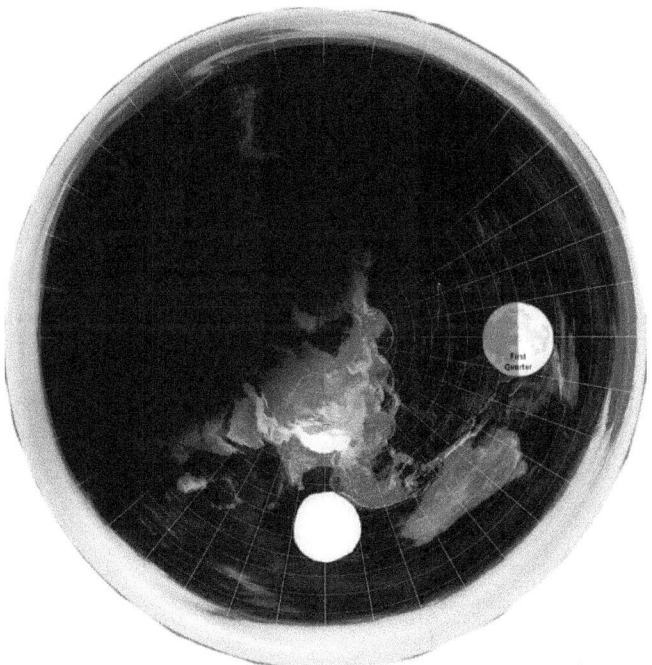

Look the at the first quarter of the moon and look her distance with the sun how it is precisely as one quarter of the clock between. Hence the name, a quarter moon.

During the year, the sun serves to count the days and the years, while the moon to count months and weeks. And both behave differently during different seasons by giving us signs in which season we stand.

---

[4] Same as 3rd reference.

As the moon goes towards waxing gibbous, so her distance with the sun increases:

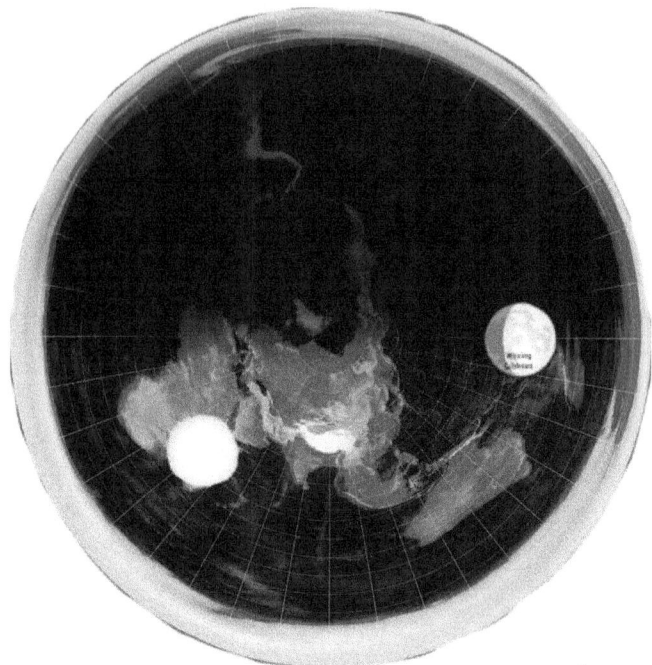

At this phase, the moon tells us that we're along the second week of the month, until the full moon.

"At **Full Moon**, the Moon is opposite to the Sun - 180 degrees away, and 12 hours behind it. So the Moon rises as the Sun is setting; it's high in the south at midnight, and it sets in the morning, at sunrise."

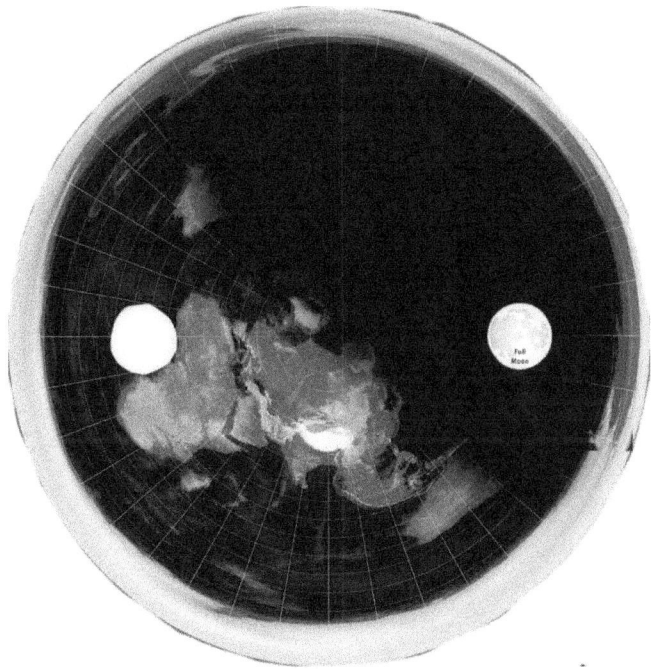

From this position the distance between the moon and the sun begins to lower and the phases of the moon lower the light from its fullest towards waning. Now the second part of the month begins. Two more weeks and the moon goes back to zero light again.

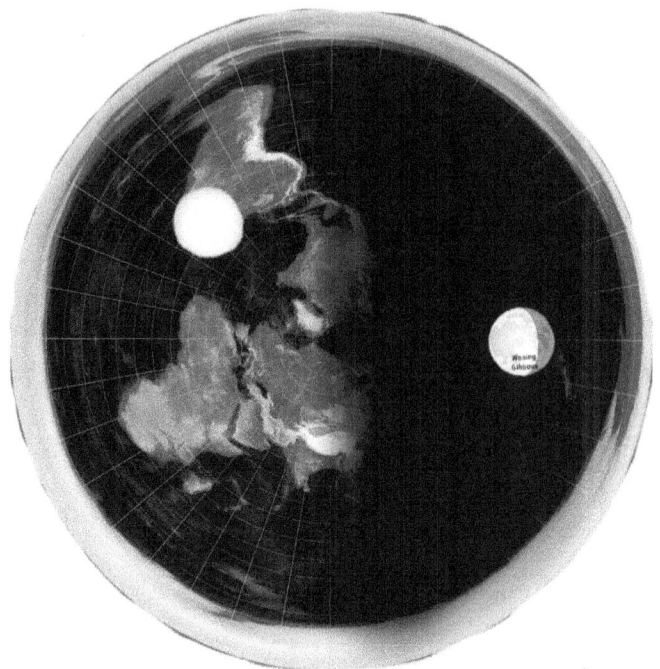

As the moon wanes towards the last quarter, it tells that we are in the third week of the month.

"By **Last Quarter**, the Moon is 270 degrees to the left of the Sun - or 90 degrees to the right of it; and it lags 18 hours behind the Sun - or it's 6 hours ahead. So it rises in the middle of the night, it's high in the south at dawn, and it sets in the middle of the day."

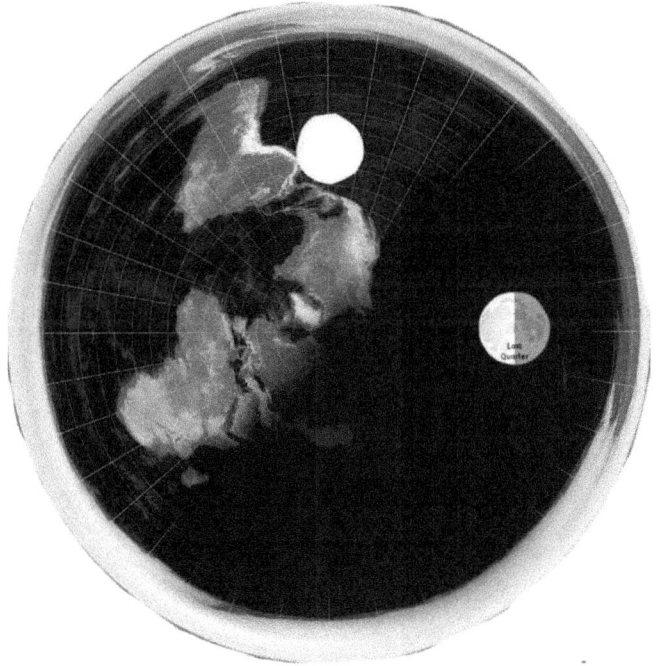

And so the sun still approaches the moon and the moon continues to wane towards the second/last crescent, marking the fourth week and that the old month is about to end and the new month is near.

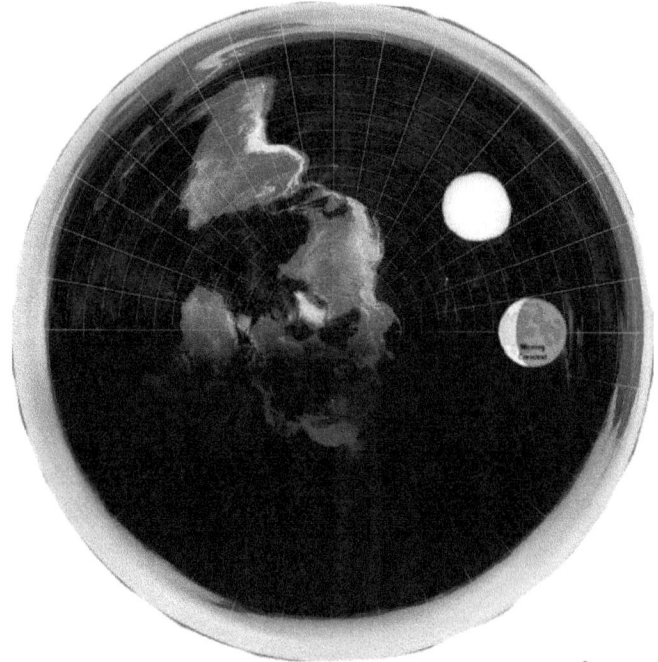

Until finally the final crescent is seen no more at dawn and the new month begins again with the new/zero moon day:

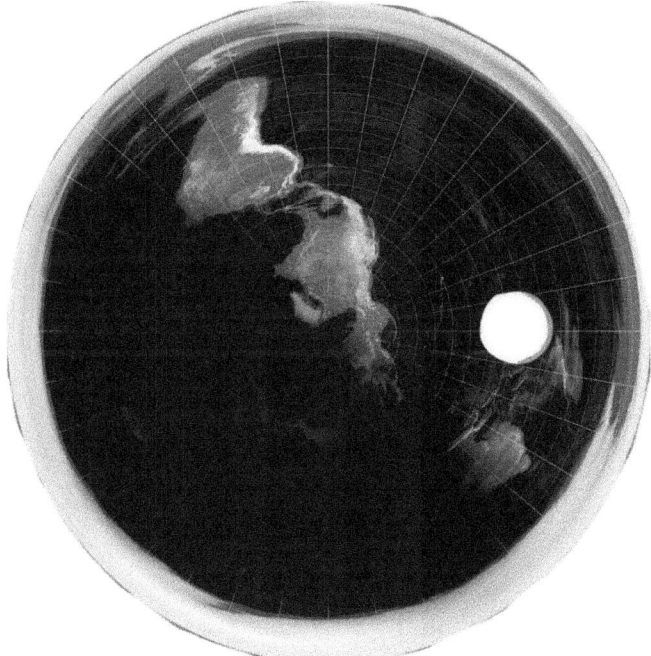

From all these representations, notice that the moon and the sun are at all time anchored relatively between each other, marking the state of time and spinning around the encircled plain earth continuously. Just like the minute and the hour hand of the clock. Minute hand being a metaphor of the sun, while the hour hand a metaphor of the moon.

New moon day is the first day of the month, a celebration and thankful day for the past month and prayerful for the upcoming one. But, the counting of the first day of the week is done after the New Moon day:

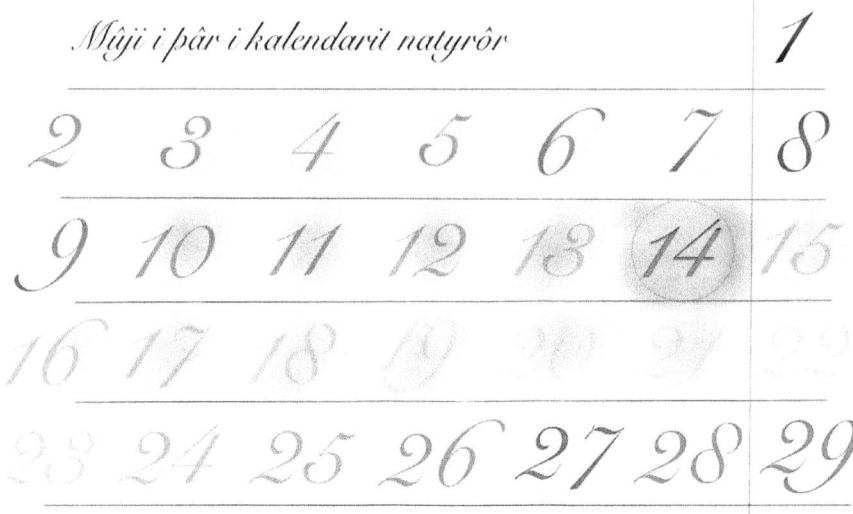

Lavdû kôft Zoti. Pronari i botnave.

Mûji i pâr i kalendarit natyrôr   *1*

2	3	4	5	6	7	8
9	10	11	12	13	(14)	15
16	17	18	19	20	21	22
23	24	25	26	27	28	29

So, based on the Genesis of Creation and the Law of Moses, Dates 1, 8, 15, 22, 29, are resting and non-working days, as Commanded by the Creator, in all the rest, shine your best.

~ Flax blooming period

There is a pointer in Torah, Exodus, chapter 9 and 12, which lets us know when the year begins.

Exodus

9:31 And the flax and the barley were smitten; for the barley was in the ear, and the flax was in bloom.

9:32 But the wheat and the spelt were not smitten; for they were not come out into ear.

12:2 This month shall be unto you the beginning of months: it shall be the first month of the year to you.

23:16 Also you shall observe the Feast of the Harvest of the first fruits of your labors from what you sow in the field; also the Feast of the Ingathering at the turn of the year when you gather in the fruit of your labors from the field.

34:22 And thou shalt observe the feast of weeks, even of the first-fruits of wheat harvest, and the feast of ingathering at the turn of the year.

http://www.historyonthenet.com/Egyptians/farming.htm

The first fruits of wheat in Egypt are ready from March to the beginning of April.

To harvest the wheat has to be in the ear first. And God is saying just before beginning the first month, that it was not. So it is not springtime. It is before spring, winter. That's when the year begins.

Then, flax was in bloom. From the stages of flax development according to the scientific article:

http://www.agriculture.gov.sk.ca/Default.aspx?DN=0aa6663b-c240-4594-8889-9f54de340c2b
2 months vegetative period. 15 - 25 days flowering period. Which makes from 2 and a half to three months.

Since early seeding brings best results, starting from October, November we arrive in December. Again time of the winter solstice.

Late seeding is not an option because the first fruits would not be at the turn of the year.

Peace and thank God, another dawn.

"And thou shalt observe the feast of weeks, of the first-fruits of wheat-harvest, and the feast of ingathering at the turn of the year." (Exodus 34:22)

What does the turn of the year mean?

is it from w.solstice to sp.equinox, or from sp.equinox to sum.solstice or from sum.solstice to f.equinox, or from f.equinox to w.solstice?

If you read the verse again, starting from the winter solstice we arrive at the spring equinox.

Because the first-fruits of wheat harvest in Egypt (and in region since it is not far) come in the beginning of April. Just at the turn of the year.

Excerpt from the article:
http://www.spectrumcommodities.com/education/commodity/statistics/wheat.html

"Egyptian wheat is produced in a very limited area along the Nile River and is delta plane. Throughout the Middle East, wheat is planting begins around the first of September and usually runs through the end of November. Harvest begins around the first of April and runs through the end of August."

It is not the middle of the year, nor the end. It is the turning of the year, because from the spring equinox the year turns from longer nights to longer days.

Also, the moon and sun in Holy Gita confirm the beginning of the Natural Year:

*24 Those who know the Supreme Brahman attain that Supreme by passing away from the world during the influence of the fiery god, in the light, at an auspicious moment of the day, **during the fortnight of the waxing moon, or during the six months when the sun travels in the north.***

25 The mystic who passes away from this world during the smoke, the night, the fortnight of the waning moon, or the six months when the sun passes to the south reaches the moon planet but again comes back.

Holy Gita is an amazing scripture to read on.

Notice, the waning moon and the second sixth months of the sun, in the south. Waning moon is towards the end of the month, the second sixth month is towards the end of the year. Which, confirms the finding that the natural year begins at the most farther point of the sun's journey in the south. In the winter solstice (for the northen part of the equator). Or more precise, the new moon ***closest*** to the longest night.

Peace, and praise be to the Lord of all worlds.

~ The Cycle of the Year

The Natural Year begins with the new moon day closest to the winter solstice. Closest, because the moon dances around the winter solstice each year, by approaching and going further from it, sometimes yet not often, matching the solstice exactly.

The sun and the moon spin around the plain earth like the minutes and the hour hand of the clock, contracting within the equatorial perimeter during the summer, and expanding beyond the equatorial circumference during the winter.

Since from above we established that the natural year is anchored with the winter solstice, the longest night, then the sun at that time is positioned at the fartherst place from the center of the earth/the north pole, and at nearest to the Antarctica.

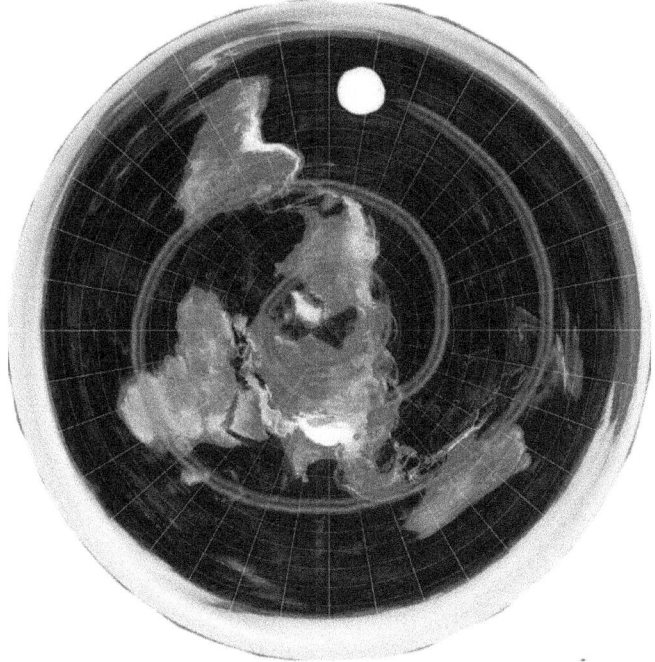

From there, it begins to lower the radius as it continuolsy spin clock-wise with the moon, thus by forming a spiral towards the centre. For six months the lumiinaries approach the north as they travel and spin, and the for the secon sixth they travel back again by increasing the radius of spinning around the centre of the earth and approaching the limit of Antarctica again, thus by completing the cycle of a year.

Samuel Rowbotham and David Wardlaw Scott are two pioneers of the 19th Century who were among the first to scientifically question the heliocentric model, thereby restoring the true form and position of the earth.

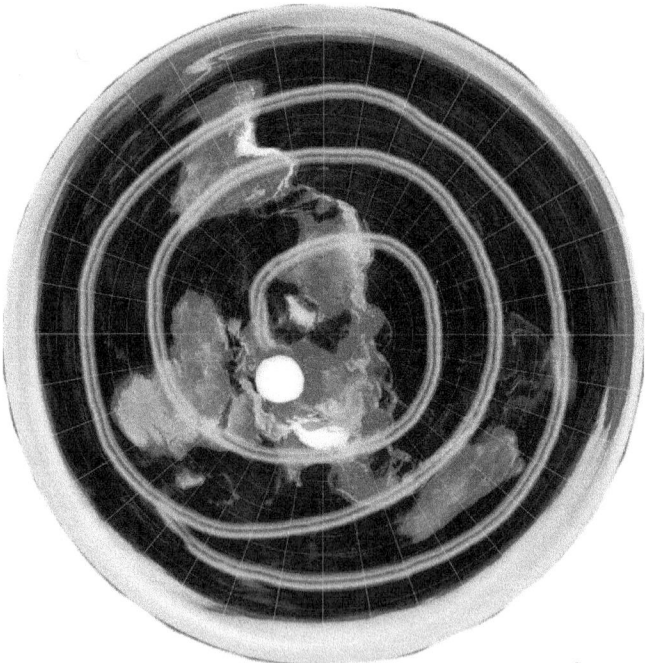

Notice that the spirals of the sun hovering above the earth are just a model and not the precise and real representation of the sun. Because, then the spiral would be much more dense, since the approaching to the north is done in much more gradual way.

~ Beautiful blend of Koran and Torah

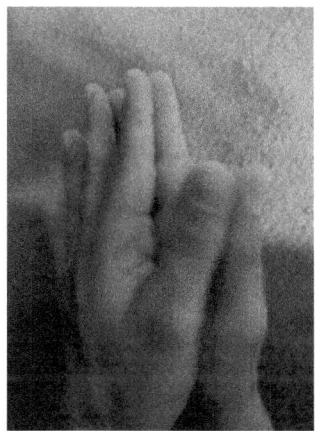

First check Koran, the chapter 89.

Beginning with the first four verses.

And simultaneously check Torah, chapter 12 of Exodus.

Notice the contrast/complement:

K 89-1: By the – dawn.

T 12-6: The congregation of Israel shall kill the lamb at – dusk.

K 89-2: And the ten nights.

T 12-3: …In the tenth day of this month…

K 89-3: And the even and the odd.

T 12-3: 10 days – even number; T 12-15: 7 days shall ye eat unleavened bread – odd number.

K 89-4: And the night when it passes.

T 12-12: For I will go through the land of Egypt in that night…. T 12-13…when I see the blood, I will pass over you… – Lord's night pass over.

Can you see the complement of each other?

How they complete each other?

Then in Koran, the same chapter continues with the stories of punishment of pharaoh and all his colleagues during the ages, executed at the same time successively.

And also in Torah, the same chapter speaks about the punishment towards the pharaoh's system.

Trying to find the truth with an isolated Scripture is like trying to paint with a single color of the rainbow.

Praise be to God. The Lord of intelligence and understanding. Peace.

Now the days are created and defined in genesis of creation. 6 working days, 1 day for rest.

Measuring the time naturally, via the moon phases per each week it is easy to discern days of rest. However, one issue has to be cleared before going on further.

~ Sabbath in Koran

2:65 You have come to know who it was among you that transgressed the Sabbath, We said to them: "Be despicable apes!"

- Decree for those who transgressed the Sabbath.

4:47 O you who have received the Book, believe in what We have sent down authenticating what is with you, before We cast down faces and turn them on their backs or curse them as the people of the Sabbath were cursed. And the will of God is always done.

- The continuum of scriptures.

4:154 And We raised the mount above them by the covenant they took, and We said to them: "Enter the passageway by crouching." And We said to them: "Do not transgress the Sabbath;" and We took from them a solemn covenant.

- Them, the group guided by Moses. The same system of the group guided by Mohammad.

7:163 And ask them about the town which was by the sea, after they had transgressed the Sabbath; their fish would come to them openly on the day of their Sabbath, and when they are not in Sabbath, they do not come to them! It is such that We afflicted them for what wickedness they were in.

- Transgressing the Sabbath makes natural missmatches.

16:124 Indeed, the Sabbath was only decreed for those who had disputed in it, and your Lord will judge between them for that in which they had disputed.

- From all these verses in Quran, regarding Sabbath, what was the decree?

- To be despicable apes, to those who transgressed it.

The System of God is One, for everyone. The Scriptures reveal that, for those who have understanding.

The Sabbath is the seventh day of Creation, confirmed by the Ten Commandments Moses received.

Most of comments in Koran, or all of them which I have seen till now, either add the unneeded (for jews and christians) in 16:124, or attribute the Sabbath to the jews.

In my light, that does not hold true.

Because all the comments I have made till now have started only from the verses the Koran offers regarding Sabbath, and from my state of knowing nothing except holy Scriptures.

Peace.

"Shabbat is primarily a day of rest and spiritual enrichment. The word – Shabbat - comes from the root Shin-Beit-Tav, meaning to cease, to end, or to rest."

In scripture is another trap of selective translation. Translation is like glowes. But the best feel is without them. Those who know do not translate. They rewrite. And the right combination of words arrives to the perfect essence.

Praise be to the God of Presence.

So "Sabbath in Koran" becomes - <u>Resting Day in Koran</u>

2:65 You have come to know who it was among you that transgressed the resting day, We said to them: "Be despicable apes!"

4:47 O you who have received the Book, believe in what We have sent down authenticating what is with you, before We cast down faces and turn them on their backs or curse them as the people of the resting day were cursed. And the will of God is always done.

4:154 And We raised the mount above them by the covenant they took, and We said to them: "Enter the passageway by crouching." And We said to them: "Do not transgress the resting day;" and We took from them a solemn covenant.

7:163 And ask them about the town which was by the sea, after they had transgressed the resting day; their fish would come to them openly on the day of their rest, and when they are not in the resting day, they do not come to them! It is such that We afflicted them for what wickedness they were in.

16:124 Indeed, the resting day was only decreed for those who had disputed in it, and your Lord will judge between them for that in which they had disputed.

Peace.

~ The Night of Decree

The night of decree is better than one thousand months. Thus says Lord in Koran. Which is that night?

Without first establishing the natural calendar of time-keeping that is impossible to find. But since in the above paragraphs we have done so, let's continue to see which that night is.

Koran 97:2 – And do you know what is the Night of Decree?

Torah, Exodus 12:12 - For I will go through the land of Egypt on that night, and will strike down all the firstborn in the land of Egypt, both man and beast; and against all the gods of Egypt I will execute judgments-- I am the LORD.

Koran 97:4 - The angels and the Spirit descend therein, by their Lord's leave, to carry out every command.

If you check the chapter 12 of Exodus, knowing that the year begins with the new moon of the winter solstice or the first new moon after the winter solstice and the fact being presented that on the fourteenth Night of the first month, God Himself executed his punishment against pharaoh, may very well be that comparing the language of God in 12:12 of Exodus and 97:4 in Koran, the night of decree is the full moon of the first month of the natural year.

Peace, and praise to the One who says - be.

May you be His bee.

By the night, the night of decree, the Lord's passover over Egypt. The final decree when Lord decided to make pharaoh let people of Israel go. The oath of God in Koran.

Exodus 11:1, ~ *Now the LORD had said to Moses, "I will bring one more plague on Pharaoh and on Egypt. After that, he will let you go from here, and when he does, he will drive you out completely.*

The final decree of the Lord before pharaoh letting people go. And Him acting at night of the date 14 of the first natural month.

Also comparing with Koran:

2:185 A month of aspiration, in which the Koran was revealed;

Aspiration for what? - For the children of Israel to be free from pharaoh.

Peace.

In Torah you find the details of Koran.

~ Another Marker - Koran and Torah

Compare the oath of God in Koran with the night of decree in Torah:

84:16 So I do swear by the redness of dusk.
84:17 And the night and what it is driven on.
84:18 And the moon when it is full.

T 12:6 and ye shall keep it unto the fourteenth day; and the whole assembly of the congregation of Israel shall kill it at dusk.
T 12:12: For I will go through the land of Egypt in that night...

The full moon of the month is the fourteenth day.

Peace.

This night, the blessed night of decree is marked with the oaths of God.

Praised be Him. The All-Wise.

~ The 19 year cycle

Notice also:

9:36 The count of the months with God is twelve months in the book of God since the day He created the heavens and the earth; four of them are restricted. This is the correct system; so do not wrong yourselves in them; and fight the polytheists collectively as they fight you collectively. And know that God is with the righteous.

9:37 Know that the use of the additional month causes an increase in rejection, for it is used by those who have rejected that they may misguide with it by making it lawful one year, and forbidden one year, so as to circumvent the count of what God has made restricted; thus they make lawful what God has made forbidden. Their evil works have been adorned for them, and God does not guide the rejecting people.

God tells us that there are 12 months since the beginning of creation, and immediately in the verse after He explains the role of the additional month.

During the cycle of 19 years, known as metonic cycle, seven years are with 13 months. That has to be kept into account in order to not loose the correlation between seasons and months.

~ Days of fasting
Compare Koran vs.Torah, Numbers.

2:185 A month of aspiration, in which the Qur'an was revealed; as a guide to the people and clarities from the guidance and the Criterion. Therefore, whoever of you witnesses the month, then let him fast therein. And whoever is ill or traveling, then the same number from different days. God wants to bring you ease and not to bring you hardship; and so that you may complete the count,

and magnify God for what He has guided you, that you may be
thankful.

9 Then the Lord said to Moses, 10 "Tell the Israelites: 'When any of you or your descendants are
unclean because of a dead body or are away on a journey, they are still to celebrate the Lord's
Passover, 11 but they are to do it on the fourteenth day of the second month at twilight. They
are to eat the lamb, together with unleavened bread and bitter herbs.

So, in the first natural month, the first ten days, beginning with the new moon day as the first day of the month, are for fasting.

This year, the new moon day of the first month, compared with pope Gregory's calendar, is the 2nd january 2014.

Praise be to God. The Lord of the worlds. Peace.

What can one observe from the nature is the fact that phases of the moon are the precision of counting the years. Also, the turns of the moon specify the week we are going thru. Up to the first half, the first week. To the full moon, the second week. To the second half, the third week. And to the zero moon, the fourth week.

~ *Comparing Gregorian, Lunar and Natural Calendar*

Gregorian calendar is a solar/seasonal calendar. It is very static and unprecise when it comes to distilling the days, because it does not take into consideration the naturalness and the purpose of the moon. Hence with this calendar your birthday is never at the same day of your birth, nor any other day of national feasts. Moreover, because of this lack of day precision, religions constantly get confused by this calendar, thus are not able to find the seventh day properly and rarely does match the correct resting day with each own proclaimed holy day.

The Lunar calendar on the other hand, does not take into account the additional month stated in Koran, thus for the same date the seasons change dramatically over the years.

The natural luni-solar calendar is the complete answer which takes into account the precision of the moon and the season of the year. It is the balanced dance of the luminaries above the still earth, which move in rotation as the hour and minute hand of the clock. That's how nature measures the time. That's how animals know when to mate. That's how the days and nights are found precisely. By the Natural Time Measurement.

~ Conclusion

So, what emerged from all this observation is:

0 – Praise be to God. The Lord of the worlds.

1) The month begins one day before the week.
2) Months begin with the new moon, with the zero light moon.
3) The new moon day is counted as – first date, predecessor of the first working day.
4) From 1st to 6th day of the week all are working days. The seventh – rest. And each other seven including new moon day.
5) Because the new moon day is the first date of the month, the first day of the week corresponds to the second date. So, dates 1, 8, 15, 22, 29 are non-working and divinely resting days. Meant from the beginning of existence to be so.
6) The beginning of the year is anchored with the winter solstice. The Natural Year begins with the new moon day closest to the winter solstice.[5]
7) The night of decree/Lord's passover is the full moon/date 14 of the first natural month, the night following the sixth day of the second week.
8) Phases of the moon and the shinning of the sun are the natural clock above the circle earth. Just like the minute hand and the hour hand of the clock. Lunar phases are the precision of the season.
9) The sun and the moon are the minute and the hour hand of the natural clock earth.
10) The natural parameters of time keeping are: dawn, dusk, daylight, nightdarkness, the moon, the sun, the seasons, the turns of seasons.

---

[5] That's for the part of the earth within the equatorial circumference, or as it is known in the ball-earth theory, the northern hemisphere.

Here is a sample of the natural year's first month:

Lavdû kôft Zoti. Pronari i botnave.

Mûji i pâr i kalendarit natyrôr 1

2	3	4	5	6	7	8
9	10	11	12	13	14	15
16	17	18	19	20	21	22
23	24	25	26	27	28	29

Night of the natural date 14 is translated as the night between 15 January and 16 January 2014.

The meaning of this month you may find in Torah, exodus 12. Koran, chapter 89, chapter 97.

Peace, and praise be to God. The Lord of existence.

Grateful,

Rinon Hoxha, rinoni@zgjidhja.com

www.zgjidhja.com,

10.12.NY;

Date 14, Month 12 of the Natural Year.

~ Outro

From all this work, one thing is to be observed carefully: *Separation is no cure for the wholennes.*

The attempt of religions to call the natural time-keeping as "islamic" or "biblical" has produced sloppy results because of the limited beginnings of thinking.

All scriptures belongs to all people. Had religions known how to judge purely with "their" proclamed scripture, surely would not end up split and grouped into circles of idelogy. Because the circles are consisted of pyramids, and one has to be really blessed by the One in order to save the own soul from the collective ignorance and stupidity. The pharaoh has drowned because he did not know the sea of Love, that Moses was preaching.

The system of God is straightness, fairness and uprightness. Nowhere in the scriptures God mentions other than, the straight path. And the straight path is not a circle. Because the circle is a curve of close-minded people because of their closed heart.

Here are the diagrams ilustrating this:

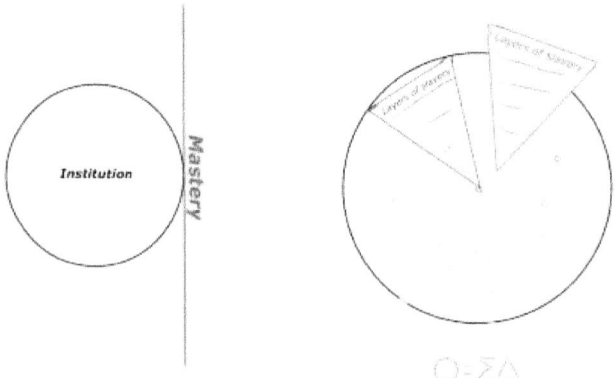

So do invite the One, the unconcievable by the mind, to set you free, totally and absolutely free from all relativity, beginning from your own content, and be yourself. Count the time with what the Lord has provided, and human did nor could nor will be able to touch and change. There is no "biblical" calendar, nor "islamic" calendar. Only the natural one.

Peace be upon you and your families, and Praise be to God. The One who turns human into a complete human being. Because _Only_ the one who has created the puzzle of existence, can find the proper place of it's pieces, to find peace and harmony of living, and wholeness of life, as meant to be.

Praise be to All-Mighty.

~ Annex 1: The Irregularity of Gregorian Calendar

The irregularity of Gregorian calendar can be observed with a simple look at the starting dates of the month.

Everyone knows that Monday is the first day of the week. The question is, why months do not begin with Monday?

Just check the months of 2014 for example: January begins with Tuesday. February, with Friday. March, Friday also. April, Monday. May, Wednesday. June begins with Saturday. July, Monday. August, Thursday. September, Sunday. October, Tuesday. November, Friday, and December, Sunday again.

So from all 12 months, only two begin with the "first" day of the week.

This fact alone concludes the irregularity of actual calendar.

And the weeks constantly get split between months. It is like saying your foot is mine, and one of my hands is yours. While naturally, the first day of the week is always the same phase of the moon.

Praise be to God. The Lord of the worlds.

Peace.

~ Annex 2: Another representation of Natural Calendar / Natural counting of Days

Since the month is counted with the phases of the moon, and knowing that the full moon is the peak of the month, makes sense that this phase be used as the mirror/anchor point of the month.

Here is the representation of the natural counting of days:

So, the full moon gets developed completely up to the date 15 of the natural calendar. Hence as the light of the moon increases from the new moon day, the numbers of days are counted upwards, while from the full moon to the zero moon, the counting is done backwards.

This numbering mirrors the light of the moon and splits the counting of dates in two phases: *before full moon and after full moon.*

For example, date 14 before full moon is the sixth day and night of the second week, while date 14 after full moon is the first day and night of the third week.

This way, the numbering is natural and from the middle of the month let's people know how many days are left to the new moon, the beginning of the next month.

Notice also for the second natural month there is a date at the end of the month numbered as date 0, which means, after the resting day of the fourth week, there is one more day and night needed for the moon to complete it's cycle around the earth, prior to beginning of the next month, hence that day is a not-counted/slider day – date zero.

Praise be to God, the Lord of existence.

Mirroring is the mastery. Check how clear and silent water mirrors the sky and the mountains. Also check how moon mirrors the sun. And above and beyond all man is made in the likeness of the Lord. Only in His Presence man is the own self and mirrors His light.

Praise be to God, the true insight.

In this kind of numbering, the full moon is the mirror of the half month before it, and the other half after it. Or it's the peak of the mountain from which one can look at the both sides of it.

Peace.

P.S. This inspiration for this kind of counting came to me first, which I did not act upon, and today it came to my wife just prior to waking up, and now I acted upon it.

Praise be to God. The Lord of guidance.

In His service,

Rinon Hoxha,

Date 4 of the 12th Month of the Natural Year.

~ Annex 3: Mathematical Approach and Finding the Proportion of Addition

From Koran:

18:25 And they remained in their cave for three hundred years, and increased by nine.

There are 15.78 cycles of 19 years in 300 years.

One 19 year cycle has 7x13 months + 12x12 months = 91+ 144 = 235 months = 19 years.

15.78 x 235 months = 3708,3 months

9:36 The count of the months with God is twelve months in the book of God the day He created the heavens and the earth;...

3708,3 months : 12 = 309.025 years

18:25 And they remained in their cave for three hundred years, and increased by nine.

...which totally make 309. This is one approach of using luni-solar counting of time to proove this verse.

Another approach is this:

If for 300 years are added 9, how many years are to be added for 19 years?

With the simple mathematics of proportion we begin:

$(300/9) = (19/x)$

$x = (19*9/300)$

19 years = 6939.55 days
300 years = 109572 days
9 years = 3287.16 days

Applying these values in the equation for x, we find the value:

x = 208.186 days

which is:

x = 7.05 months

So, if for 300 years are added 9, for 19 years are added 7 months.

Which corresponds to the natural 19 year cycle where 7 years are with 13 months.

Hence, 7 months added for 19 years.

With this the natural time counting is also mathematically concluded.

Peace, and Praise be to God. The Lord of the worlds.

R.

~ *Annex 4: Finding Your Natural Birth-Day*

Finding your natural birth day requires a few simple steps.

1. Take a lunar calendar with Gregorian dates in it. (Ex. http://www.calendar-12.com/).
2. In the year you were born, notice the winter solstice before your birth day.
3. Find the zero/new moon ***Nearest*** to the winter solstice.
4. Count the months from new moon to another beginning from the new moon at point **3.**
5. Find your birth day, and the phase and the age of the moon for that day (Ex. http://www.calendar-12.com/moon_calendar/1980/march).

The day of the moon of a certain month is your natural birth-day.

For instance, my birthday according to Gregorian calendar is 29.March.1980

Walking through the above steps 1-5, I find the date 13, 4th Month of the Natural Year.

So my true birthday is: ***13.04.NY.***

Now every year, I check for the date 13 of the 4th Natural Month of the Year and thus find my precise birth-day. Of course this unchangeable date of the Natural time-keeping is different for each year of the Gregorian calendar. Thus, for this year my birthday is 13 April 2014. Whereas for the 2013, my birthday was on March 24th.

That's why Gregorian calendar is very unreliable and unprecise when it comes to distilling the days. Your natural and true birthday is the day the season and the phase of the moon match.

One question is, what if your birthday falls on the 13th month of an embolismic year? [1]

The answer is, you find your birthday in the 12th month of each other year of 19 year cycle.

Peace and don't be surprised when your family looks at you weird when you say when is your birthday this year ☺

Praise be to God. The Lord of the worlds.

Gratefully,

R.

1) *The embolismic year is the year where there are 13 months. 7 such years are inside the 19 year cycle.*

www.ebl-ks.org;   rinonh@ebl-ks.org;

Comparison between similarities in Torah and Koran

9:4 And Moses spoke unto the children of Israel, that they should keep the passover.

9:5 And they kept the passover in the first month, on the fourteenth day of the month, at dusk, in the wilderness of Sinai; according to all that the Lord commanded Moses, so did the children of Israel.

9:6 But there were certain men, who were unclean by the dead body of a man, so that they could not keep the passover on that day; and they came before Moses and before Aaron on that day.

9:7 And those men said unto him: 'We are unclean by the dead body of a man; wherefore are we to be kept back, so as not to bring the offering of the Lord in its appointed season among the children of Israel?'

9:8 And Moses said unto them: 'Stay ye, that I may hear what the Lord will command concerning you.'

9:9 And the Lord spoke unto Moses, saying:

9:10 'Speak unto the children of Israel, saying: **If any man of you or of your generations shall be unclean by reason of a dead body, or be in a journey afar off**, yet he shall keep the passover unto the Lord;

9:11 **in the second month on the fourteenth day** at dusk they shall keep it; they shall eat it with unleavened bread and bitter herbs;

2:183 O you who believe, fasting has been decreed for you as it was decreed for those before you, perhaps you may be righteous.

2:184 A few number of days; however, **if any of you is ill or traveling, then the same number from different days**; and as for those who can do so but with difficulty, they may redeem by feeding the needy. And whoever does good voluntarily, then it is better for him. And if you fast it is better for you if only you knew.

Education Beyond
Limits

www.ebl-ks.org; 					rinonh@ebl-ks.org;

2:185 A month of aspiration, in which the Qur'an was revealed; as a guide to the people and clarities from the guidance and the Criterion. Therefore, whoever of you witnesses the month, then let him fast therein. **And whoever is ill or traveling, then the same number from different days.** God wants to bring you ease and not to bring you hardship; and so that you may complete the count, and magnify God for what He has guided you, that you may be thankful.

This is another pointer when the fasting is. In the first month of the Natural Year. The time when God executed judgments (Night of Decree) against Pharaoh's System.

Notice the nature of language, the solutions when one is traveling, ill or unclean – the same number from different days, the same day of the next month.

See?

The Year is to begin clear by fasting the first 10 days, continuing with the Passover Sacrifice and from the dusk of the 14 day is the Great Night of Decree, when Lord has Executed Judgments.

Then from the Next Day which is the second resting day of the first Natural Month, is to be consumed unleavened bread, meaning, without memories from the past, clean.

So, 10 days fast, Sacrifice, Great Night, 7 Days Unleavened Bread.

This is the Prescription of Beginning the New Year by Holy Scriptures.

Peace and Praise be to God, the Lord of all that is.

R.

www.ebl-ks.org;   rinonh@ebl-ks.org;

Second Mirror Part of the Year

By Rinon Hoxha

If you check Torah, there you may find a similarity between the longest feasts the Lord decrees. One is the 7 days of Unleavened bread beginning at the Great Night of Decree, and the second is the 7 days of Tabernacles beginning also at the full moon of the 7th Month.

The other similarity between these two feasting intervals is the 10th day of the 1st Month and the 10th day of the 7th month.

On 10th day of the 1st Month, people are commanded to select a good he-lamb/goat without blemish of one year and keep it until the dusk of the Night of Decree for sacrifice.

Night of Decree is among the longest nights of the year for the Northern Hemisphere, where it was decreed.

On the other hand, the 10th Day of the 7th Month, is the Day of Atonement, a Great Day of Forgiveness set by the Lord. As the Night of Decree falls among the longest nights of the year, just after the winter solstice, complementary does the Great Day of Atonement which is immediately after the Summer Solstice, among the longest days of the year.

So both those seven days feasts are complementary and mirror of each-other.

Now, let's check another indicator that points the difference between starting the Calendar Naturally anchored with the Winter Solstice vs. the usual Spring Equinox via today's jewish tradition.

First let's examine these verses from Torah, Leviticus 23:39 and 23:40

23:39
Howbeit on the fifteenth day of the seventh month, when ye have gathered in the fruits of the land, ye shall keep the feast of the Lord seven days; on the first day shall be a solemn rest, and on the eighth day shall be a solemn rest.

Education Beyond
Limits

www.ebl-ks.org;   rinonh@ebl-ks.org;

23:40
And ye shall take you on the first day the fruit of goodly trees, branches of palm-trees, and boughs of thick trees, and willows of the brook, and ye shall rejoice before the Lord your God seven days.

By checking these observations and the time of the season when fruits are gathered and are fresh:

http://www.bible-history.com/geography/seasons_months_israel.html

as well as,

https://www.gci.org/law/festivals/harvest

One may see that despite rationalization from these articles, the tables and the seasons in which most fruits correspond have to do and are in direct correlation with starting of the New Year anchored with the Winter Solstice, and Not with the Spring Equinox.

Completing this with the greatest length of the day during the year, the 7th Natural Month, in which the Day of Atonement is, concludes the other side of the year.

So, Great Night of Decree / Lord's Passover – Among the Longest Nights of the Year, corresponds to the Great Day of Atonement – Among the Longest Days of the Year.

The 7 Days of Unleavened Bread – Start at the 15th day – The full Moon, correspond with the 7 Days of Tabernacles – which also. begin by the – Full Moon, Second Resting Day of the Month.

The only Difference is the Season: One Winter, Second Summer. Also, these are the only two seasons mentioned throughout Holy Scriptures. Winter and Summer.

106:2 The way they cherish the journey of the winter and summer. ~ Koran.

Peace, and Praised be All-Mighty.

Rinon Hoxha,

18th Night of the 1st Month of the 2nd New Natural Year.   *(8.Januar.2015)*

Second Mirror Part of the Year © Rinon Hoxha.

www.ebl-ks.org;   rinonh@ebl-ks.org;

The Great Day of Rebirth

by Rinon Hoxha

Learning to Judge with the Word of God requires learning not to judge with the tradition of human. Interesting fact is that each today's tradition, be Jewish, Christian or Muslim, all have failed miserably in holding onto God and His Word, alone.

Each group has invented the own idols, and today worship them diligently while thinking that are doing the right thing, because of the own conviction.

Nevertheless, thanking God that I am neither Jew nor Arab and being raised in a tolerant multi-cultural/religious environment where religion was never a reason of fight, had the chance to approach all Holy Scriptures open-heartedly and from a neutral standpoint.

This in itself has opened and continues to open-up myriads of doors that others cannot even consider that exist.

Tradition is an Albanian word - T'râ-Dita - which means – Hit of Light. This is certainly true if the light bypasses all the forms through. But the problem of most traditions, if not all, is clinging on the forms of idolatry. And any idolatry is a believed form which manifests itself via eyesight.

For instance, today's Muslim feasts known as Mubareq (the feast at the end of todays 30 day fasting) or the other feast where most Muslims make sacrifice for the remembrance of Abraham and his own personal test of life and his son, tell a lot how astray today Muslim community has gone.

The first reason is because of un-natural time keeping, thus movement of "Fasting Month", whereas in Koran, the Lord mentions only …. a few number of days.

The second reason is the intent of slaughtering animals like Abraham in his own time did. Moreover, this also misses the time when this should be done and the total missing of the point why it should be done. The only explanation of – Why – to do a sacrifice for the Lord is found in Torah, at the dusk of the Night of Decree, an ordinance to be kept

Education Beyond
Limits

www.ebl-ks.org;   rinonh@ebl-ks.org;

forever – For the Remembrance of the Lord. And not how today's Muslims do, for the challenge of Abraham.

You can see though, again, how idolatry kills the slaughterer instead of the sacrificed animal.

This was for today's Muslim Community, who has created almost uncountable number of sects and sub-sects, each being driven by the own idols.

Now onto the Jewish Group who mostly keep themselves as "the chosen ones", while forgetting that they were among the first to step in the role of Pharaoh and idolatry immediately after crossing the Red Sea. Also, for missing a lot of points and deliberately choosing not to accept Koran which confirms the Torah that have let behind and brought in-front the Talmud to go after what has no flow.

Then, the Christians, who are almost a mirror of today's Muslims with the number of sects, and an open idolatry towards prophet Jesus. A totally controlled mindset from the centres they belong, and totally closed to see beyond their ideology. Thus, not accepting that there cannot be a complete book without the last chapter of Inspiration.

So, in all this mix of "True Paths" one has to let go the path of tradition and make to the Scriptures no addition. Tradition is the leaven which prevents the heaven. Muslims, Jews and Catholics – have to – let go what has become the barrier towards the God.

He has sent Prophets in many languages, and most people since then have been stuck in the language they received the Scripture in. The truth needs no language. It only needs the deepness of the heart and understanding art.

There is a simple reason why all these communities fail constantly: Because no one accepts the Scripture of the other. Why? – Because – Expecting the truth based on like is losing the likeness of the Lord.

And All we have begun in Genesis. But the hell we have created for ourselves on this very earth.

Let's check again the Two basic Scriptures, the left and the right hand of Understanding:

Education Beyond
Limits

www.ebl-ks.org;   rinonh@ebl-ks.org;

Koran:

20:102 *The Day* **the horn is blown**, *and We gather the criminals on that Day white eyed.*

20:103 *They whisper among themselves: "You have only been away for a period of ten days."*

20:104 *We are fully aware of what they say, for the best among them will say: "No, you have only been away for* **a day**."

Torah, Leviticus:

23:23 *Tell the Israelites, 'In the seventh month,* **on the first day** *of the month, you must have a complete rest, a memorial announced by* **loud horn blasts**, *a holy assembly.*

23:27 *In addition,* **the tenth day** *of this seventh month is a special day for* **the payment for sins.** *There will be a holy assembly. Humble yourselves, and bring the LORD a sacrifice by fire.*

Unlike what Campbell says, I point that Holy Scriptures are not just metaphors but direct pointers to the truth and the mirror of the truth as well. All Relativity is a metaphor of Reality. To get to the Real part of Understanding the reader has to open up the own self and let go the own sins. There are specific times of prayer and specific Days when the Lord Himself has marked those days.

One is the Great-Day of Purification, a chance to purify the own soul from the trashes of the own self.

If you check the above verses of Koran and Torah, you will see that each year, the chance of becoming at one with the One is given to everyone with His blessing.

All you have to do is let the Scriptures sing the One song of the One.

This is what being – People of the Book – is. All Scriptures are the Same Book of the Same Lord.

The Great Day of Rebirth © Rinon Hoxha.

Education Beyond
Limits

www.ebl-ks.org;   rinonh@ebl-ks.org;

Notice those verses yourself and relate the meanings by color and boldness.

Praise be to Lord of the forgiveness.

Peace,

Rinon Hoxha.

18th Night of the 1st Month of the 2nd New Natural Year.   *(8.Januar.2015)*

www.ebl-ks.org;   rinonh@ebl-ks.org;

Calendation: Months of Celebration and Feasts

by Rinon Hoxha

The last clear-up about the feasts via Scriptural terms[1] are as follows:

1st Month and the 7th Month are a mirror of each other. This means, around each solstice, respectively the Winter and Summer solstice.

Let's see the graphical representation of both of them:

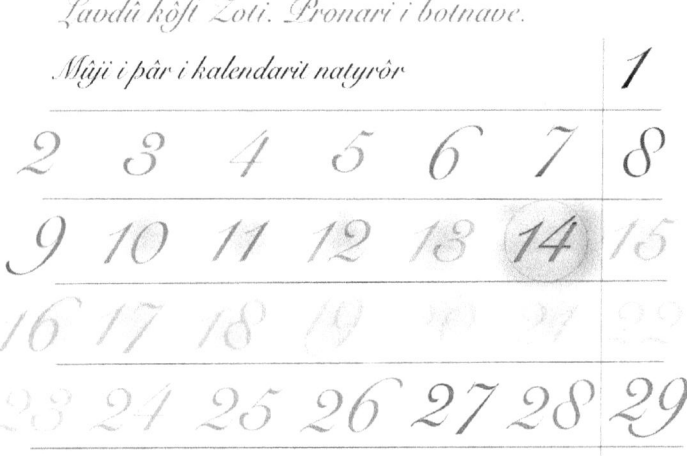

In the First Month, the blessings are as follow:

---

[1] Conclusions are deriven from Koran and Torah.

Education Beyond
Limits

www.ebl-ks.org;   rinonh@ebl-ks.org;

Dates 1-10 Fasting.

Dates 10-14 Choosing the right animal and Sacrificing for the Lord.

From the Dusk of the 14th Day to the Dawn of the 15th Day is the Blessed Night of Decree / Lord's Passover.

Dates 15-21 The 7 Days Feast of Unleavened bread.

This concludes the Celebrations of the First Month of the Natural Year, which begins with the New Moon/Zero moon after the Winter Solstice.

The Seventh Month is on the other side of the Natural Year. It is anchored nearest to the summer solstice. Again, similarly with the First Month:

Lavdù kôft Zoti. Pronari i botnave.

Muji i Shtát i kalendarit natyrôr   1

2	3	4	5	6	7	8
9	10	11	12	13	14	15
16	17	18	19	20	21	22
23	24	25	26	27	28	29

Calendation: Months of Celebration and Feasts © Rinon Hoxha.

Education Beyond
Limits

www.ebl-ks.org;   rinonh@ebl-ks.org;

Date 1 – Feast of Horns.

Date 10 – Day of Atonement – The Great Day of Forgiveness.

Dates 15-21 The 7 Days Feast of Tabernacles.

Date 29 – Feast of Weeks.

The only feast that is alone from these two blessed months is the Feast of the First-fruits as a gratitude for the first-fruits that show up in Spring.

So, all celebrations/observances are in only these two natural months: The 1st and the 7th.

One Celebration Month in the Beginning of Winter Season, the Second in the beginning of the Summer season, for the Northen Hemisphere of the Earth.

Or

For the Southern Hemisphere is the vice-versa, the First Month is in the Summer Season and the Second Celebration Month in the Winter Season.

With this, dear readers, is finished and completed the Natural Time Keeping calendar.

It is restored according to Holy Scriptures, as meant by All-Mighty.

Peace and Praise be to God, the Lord of the worlds.

Rinon Hoxha.

20th Night of the 1st Month in the 2nd New Natural Year.   *(10.Januar.2015)*

Education Beyond Limits

The Final Revelation about the Beginning of the Moon-th

Since re-establishing that the earth is encircled plain and the luminaries of the sun and moon are spinning clockwise complementary to each other, each according to their own light during day and night, it has been a joy for me to re-check again the beginning of the month in the new light which came from light upon light.

Many communities historically have been relying in the crescent phase of the moon, or in the new moon, when it is seen most thin, or even the new moon without seeing it but knowing from the Heliocentric Model when the conjuction (the straight position of the moon between the supposedly ball earth and the sun takes place) so that that point in space is taken as the basis for establishing the beginning of the month.

However, as God wants the life to be a long living and learning journey, from deeply meditating again upon the luminaries that flow and float in the own heaven upon the flat earth, I have come to this conclusion:

The month begins with the zero moon. Empty moon, no light. Ready for the next insight.

Being cleared up from the hypnosis of the heliocentric model, when you look with the heaven's eyes here's what you get, citing Koran:

36:39 And the moon We have measured it to appear in stages, until it returns like an old palm trunk.

From this verse and from night-to-night experience we may attest that the moon appears in stages. From a crescent moon after dusk, which appears in the beginning days of the month, to the crescent moon of dawn (old palm trunk), which appears at the ending of the month.

Now in this position there is a gap. And that is the time where the moon does not appear at all, or it cannot be seen. It can go for 1-2 days.

Education Beyond
Limits

www.ebl-ks.org;   rinonh@ebl-ks.org;

The question then becomes: Which part of the unseen moon shall we take as the beginning of the month? The vanished moon immediately after the last seen crescent (like old palm trunk at dawn) or the thin crescent known in christian and mohammedan cultures as the New Moon?[1]

The answer to this question we may carefully find from exploring both sides of the verse 2:189

2:189 They ask you regarding the crescent moons, say: "They are a timing mechanism for the people and the Pilgrimage." And piety is not that you would enter a home from its back, but piety is whoever is righteous and comes to the homes through their main doors. And be aware of God that you may succeed.

The first part of the verse notes that the Prophet was asked about the crescent moon-s, or crescents. The inspirational answer from the Divine is: "They are timing mechanism for the people and the Pilgrimage".

Now at the second part of the verse it becomes interesting: *And piety is not that you would enter a home from its back, but piety is whoever is righteous and comes to the homes through their main doors. And be aware of God that you may succeed.*

At first looking superficially it may seem that this part has nothing to do with timing, but consider:

There are two types of crescents: The dawn crescents at the end of the month and the dusk crescents at the beginning of the month.

These are markers, or timing mechanisms.

Now look again at the previous verse: *And the moon We have measured it to appear in stages, until it returns like an old palm trunk.*

So, that's her final phase, the last crescent of dawn, the old palm trunk. Which means, anything, or nothing better to say since it cannot be seen anymore is the new beginning. At the zero moon. Empty.

---

[1] Note the words New Moon and Crescent Moon are used interchangeably in both cultures.

Education Beyond
Limits

www.ebl-ks.org;   rinonh@ebl-ks.org;

Return to the the piety: *And piety is not that you would enter a home from its back, but piety is whoever is righteous and comes to the homes through their main doors. And be aware of God that you may succeed.*

In this case, since this is the part of the verse about crescents, it can be easily interpreted like this:

Home, to begin, purely, in piety, zero, empty, to begin like that, you have to go through the main door-s.

Back door being the last crescent of the dawn, and the first crescent of the dusk. Because, both crescents are the back of the moon.

Look at both crescents in the figures below, respectively:

… but piety is whoever is righteous and comes to the homes through their main door-s.

So, what's through the main doors? – Empty moon.

Thus, the month begins, one day **after** the last seen crescent in the dawn. So, when the crescent of the dawn cannot be seen, that's the New Moon day, which begins with the empty,

www.ebl-ks.org;   rinonh@ebl-ks.org;

zero moon. That's the day to contemplate upon the previous month, give thanks, and pray for the upcoming one.

That's the - Date 1 - of the Natural Month.

Then, after seven days, in the Date 8 is the First Resting Day. Succesively, seven after seven, are the next resting days, in the dates: 15th, 22nd and 29th when the crescent of the dawn indicates the ending of the Natural Month.

Praise be to God above all domes of heaven, for deepest and highest knowledge.

Peace,

Rinon Hoxha.

Date 21 of the 3rd Month, the Second Natural Year.

(11.03.2015).

www.ingramcontent.com/pod-product-compliance
Lightning Source LLC
Chambersburg PA
CBHW071003180526
45168CB00003B/1265